COAL MINING IN BRITAIN

Richard Hayman

Published in Great Britain in 2016 by Shire
Publications (part of Bloomsbury Publishing Plc),
Kemp House, Chawley Park, Oxford OX2 9PH, UK

29 Earlsfort Terrace, Dublin 2, Ireland
1385 Broadway, 5th Flr, New York, NY 10018, USA

Email: shire@bloomsbury.com
www.shirebooks.co.uk

© 2016 Richard Hayman.

A CIP catalogue record for this book is available from
the British Library.

Shire Library no. 836.   ISBN-13: 978 1 78442 120 5

PDF eBook ISBN: 978 1 78442 121 2

ePub ISBN: 978 1 78442 122 9

Richard Hayman has asserted his right under the
Copyright, Designs and Patents Act, 1988, to be
identified as the author of this book.

Typeset in Garamond Pro and Gill Sans.

Printed and bound in India by Replika Press Private Ltd.

21 22 23 24 25   10 9 8 7 6 5 4 3 2

COVER IMAGE
Front cover: The Big Pit National Coal Museum at
Blenaevon (© The Photolibrary Wales / Alamy Stock
Photo); Back cover: An illuminated Davy Lamp.

TITLE PAGE IMAGE
Screening in the nineteenth century is shown in an
advert from the *Colliery Guardian*. Coal was tipped
from tubs on to the screens and fed into railway
wagons beneath them.

CONTENTS PAGE IMAGE
As underground working became more extensive in
the early twentieth century, manriding cars powered
by electricity were introduced to move miners quickly
to the coal face.

ACKNOWLEDGEMENTS
Permission to reproduce illustrations has been given
by the following: Archive Images, page 54; Beamish
Museum, pages 39, 40 (bottom), 44, 55, 57 (top), 58;
Durham Miners Association, page 57 (bottom); Elton
Collection, Ironbridge Gorge Museum Trust, page 14;
istock, page 34; Shutterstock, pages 4, 20, 24, 60; UK
Coal, page 31 (top); www.picturethepast.org.uk, pages
40 (top), 45, 53; Yale Center for British Art, Paul
Mellon Collection, page 22.

Shire Publications is supporting the Woodland Trust, the UK's leading woodland conservation charity, by funding the dedication of trees.

CONTENTS

BLACK GOLD   4

BELL PITS AND HORSE WHIMS   8

DEEP MINING   16

GOING UNDERGROUND   34

THE PIT VILLAGE   48

PLACES TO VISIT   59

FURTHER READING   63

INDEX   64

BLACK GOLD

'THE MATERIAL SOURCE of the energy of the country – the universal aid – the factor in everything we do'. That was how the Victorian economist W. Stanley Jevons described the coal industry in 1865. Coal had been mined for centuries but in the eighteenth and nineteenth centuries the coalfields became Britain's industrial heartlands. Britain became the first nation to base its economic civilisation on mineral fuel and rose to be the world's largest economy. How this came about is quite simple. Coal became the fuel of the iron industry, by far the most important of the metals trades in Britain. In turn the iron industry gave us fuel-hungry steam engines, and the railways and ships by which coal could be distributed cheaply at home and abroad.

The decline of the coal industry in the second half of the twentieth century was rapid. When it was formed in 1947 the National Coal Board (NCB) was responsible for over 1,500 collieries. Coal was central to the vision of a bright future for Britain and its industry, but it was soon eclipsed by oil and very quickly it has been consigned to the past, with a reputation as a dirty fuel and a significant contributor to global climate change.

Clearance of collieries and landscaping of old spoil tips has removed the industry from its former dominant presence in the landscape, so that in parts of the coalfields the only reminders of the former industry are subtle ones such as old miners' institutes or the sterile landscape of re-graded

Opposite: The engine house and pit-head buildings at the former Barnsley Main Colliery in South Yorkshire, which closed in 1991, are among the few colliery buildings to escape the rapid clearance of coal-industry sites in the late twentieth century.

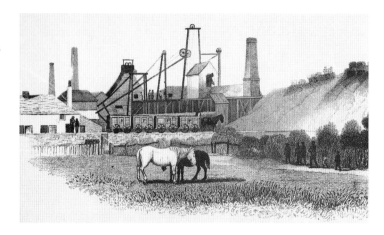

Wideopen Colliery, Northumberland, viewed here in 1844 by Thomas Hair. Until the nineteenth century the northeast was the dominant region of coal production.

spoil tips. Other aspects of coal-mining life have disappeared completely. The special language of the 'goaf' (the space where coal had been cut from), the 'rolley-way' (or underground tramway) and the 'dib-hole' (the sump for collecting water at the bottom of a shaft) no longer means anything. The skills of the face workers – known mainly as hewers, but as haggers in Cumberland, pikemen in Shropshire and getters in Yorkshire – have also vanished, a reminder that industrial decline brings with it cultural as well as economic losses.

Nevertheless, there is still a rich heritage of coal mining in Britain, albeit one that engenders mixed feelings. It produced a skilled workforce vital to the nation's interest, but in the process miners endured hard times and appalling sufferings. Outside of the coalfields the miner has always been something of a mythical figure, the consequence of which is that society has often viewed him (occasionally her) unsympathetically. In modern times this has been partly a political response, since coal miners became a recognised force in the nation's political life, forming a vociferous and often militant lobby for better working conditions and a decent standard of living. But miners have always seemed like a community apart, a fact determined by the nature of the work and by the geographical locations in which the industry was confined.

Chatterley Whitfield, near Stoke-on-Trent, is the most extensive of the surviving collieries in Britain, although coal was last brought to the surface here in 1976.

The dangers of coal mining have cast a shadow over the history of the coalfields. The worst British peacetime disasters have happened in the coal industry. Disasters can be measured in numbers, on which basis the worst event occurred in 1913 when 439 men and boys from the Universal Steam Colliery at Senghenydd, in South Wales, were killed by an underground explosion. Or it can be measured on a scale of unimaginable horror, in which case nothing could eclipse the disaster at Aberfan near Merthyr Tydfil in 1966, when 144 people, 116 of them school children, perished under a collapsed colliery tip. In the face of these tragedies the coalfields produced communities with a distinct politics and culture, perhaps best expressed now in the still-flourishing Durham Miners' Gala, who were proud of their contribution to Britain's economic wellbeing.

'The last dram of coal' sums up the pride in mining communities and their sense of loss at its decline.

BELL PITS AND HORSE WHIMS

COAL IS A mineral fuel. It belongs to the geological period known as the Carboniferous, and was laid down between 306 and 313 million years ago, when Britain enjoyed a tropical climate. It was formed from thick layers of peat in river deltas that slowly subsided under their own weight and were eventually compressed to form coal – it took a layer of peat 10 metres thick to produce a coal seam 1 metre thick. Often the deposition of peat was interrupted by layers of sediment on which another phase of peat would later form, which is why coal is found as seams in sedimentary rocks. Subsequently the individual seams, known as coal measures, were subject to various geological processes, like uplifting, faulting and folding, which have had two major consequences. Coal seams outcrop on the surface in all of Britain's coalfields, except in Kent, but the unpredictable nature of coal seams means that mining operations have often been confounded by geological faults.

Coal seams are formed in sedimentary layers and have been subject to various geological processes such as uplifting and faulting.

Coal is not uniform, so its carbon content, and therefore its calorific value, varies. The coalfields of Britain have yielded mainly hard bituminous coal, which has a carbon content of between 86 and 88 per cent. Known also as steam coal or coking coal, its main uses in the industrial revolution were for

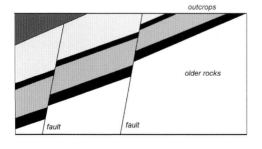

outcrops

older rocks

fault   fault

engine boilers and for conversion to coke for the iron and steel industry. In the twentieth century it was also used for electricity generation. Oil and gas have substantially replaced steam coal, but no substitute has yet been found for coking coal. An even harder coal is anthracite, which has a carbon content of about 94 per cent and is an ideal domestic fuel because it emits little smoke when burned. The principal source of it in Britain was the western part of the South Wales coalfield. Lower grades of coal, known as brown coal or lignite, are not mined in Britain.

Coal is extracted from the ground in four main ways, depending upon geological conditions and its proximity to the surface. The earliest and many of the latest coal workings exploit coal on the surface. Coal pits, or collieries, are usually vertical shafts, from the base of which seams of coal are exploited. Occasionally it is easier to exploit the coal by driving underground at a steep angle; these mines are known as drift mines. Where the terrain is hilly it is sometimes possible to tunnel straight into the hillside, with a very slight uphill gradient to allow water to drain away naturally. These adits, or 'headings', are found only at small-scale mines in Britain.

Coal is most useful when it can be brought to the surface in manageable lumps, and so the task of the colliers was to bring the coal to the surface with as little waste as possible. Waste included other stone, known as 'dirt', and smaller pieces of coal, known as 'slack'. There was only a limited demand for slack, mainly for steam-engine boilers and salt pans, until the twentieth century, when lower grades of coal found new uses as technology found ways to make the most of dwindling coal reserves. Some furnaces and power stations now burn what is little more than coal dust.

Coal was certainly dug in Roman times, since evidence for it has been revealed in archaeological excavations. It has been found at some of the military stations on Hadrian's Wall

At Benwell Colliery, near Newcastle, coal was loaded from wagons directly onto vessels on the Tyne. Coastal trade was the foundation of the region's industrial prosperity.

and seems to have been used at the Temple of Minerva in Bath, to fuel the sacred flame. Coal mining is not mentioned in the Domesday Book of 1086, however, although the industry revived and grew in the Middle Ages. By the 1500s nearly all of Britain's coalfields were exploited to some extent. Demand grew as the metals trades developed – coppersmiths, pewterers, armorers and gunsmiths all used it. Coal was also used by lime burners, and in the heating of pans to extract salt from sea water. By the seventeenth century its use was extended to glass manufacture, the production of bricks and tiles and earthenwares. By burning off impurities to make coke in a similar manner to that by which charcoal was derived from wood, it was also used in the brewing industry for drying malt. The main market for coal, however, was for domestic use.

In this period coal mining prospered most in those areas where coal could be mined and transported cheaply to its markets. Coal is heavy and bulky in relation to its value, so its usefulness always depended on an effective and cheap means of transportation. Until the eighteenth century the predominant coalfield was in Northumberland and Durham, where over one million tons of coal are estimated to have

been mined in the
1680s, accounting for
over 40 per cent of the
nation's output. The coal
industry was aided by
access to the ports of the
Wear and Tyne, from
where coal was exported,
mainly as coastal traffic

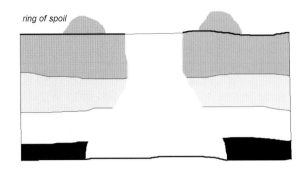

ring of spoil

to London. Other coalfields found their own markets.
Coal from northwest England was shipped to Ireland. The
River Severn cuts through the end of the East Shropshire
coalfield, which allowed coal to be shipped by river to towns
such as Shrewsbury, Worcester and Gloucester. Likewise the
River Wye allowed Forest of Dean coal to be sold up-river
in places such as Hereford and downstream to Bristol and
West Country ports. This pattern would not change until
the eighteenth century. Increasing use of coal as an industrial
fuel saw the rise of the Midlands coalfields, followed by the
rise of the Yorkshire and South Wales coalfields in the late
eighteenth and into the nineteenth century.

Coal was initially dug at or close to the surface, and sea
coals – lumps of coal washed up on to beaches from coastal
outcrops – were exploited in Northumberland and eastern
Scotland. As soon as these reserves dwindled, coal could only
be won from greater depths. It was initially dug from small
bell pits, but they were seldom more than 5 yards deep. Few
of them now survive because of the ease with which such areas
could be fully worked out by mechanical opencast methods
in the twentieth century. Coal was dug out from the bottom
of the shaft in all directions as far as safety would allow, and
so the workings had a characteristic bell shape, with a ring of
spoil around the surface of the pit. Coal could be shovelled
out of the pit manually or, when it became too deep, was
raised by a simple windlass or whim.

Bell pits could
be extended
outwards
only as far as
was possible
without the roof
collapsing. To
exploit a single
seam, therefore,
it was necessary
to sink multiple
pits.

This plan shows a coal seam in the northeast where the pillar-and-stall method remained in use in the nineteenth century. Compared with the longwall method it was less efficient as it left a grid of large unworked pillars of coal that supported the roof.

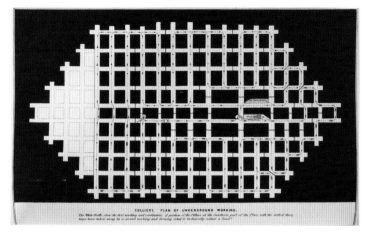

COLLIERY,   PLAN OF UNDERGROUND WORKING.

In this idealised view of a Staffordshire pit, coal is being dug by the pillar-and-stall method, and hauled up by rope.

Working in bell pits was inefficient because multiple pits had to be sunk from the surface to exploit a single seam. Greater efficiency was gained by working out from a single shaft, in a technique that could be applied to workings of any depth. By the eighteenth century shafts could be up to 70 metres deep and shaft sinking was a specialist skill separate from those of the collier. To prevent the collapse of the workings the 'pillar-and-stall' method was developed, in which 'stalls' were dug out, leaving 'pillars' of coal at intervals to support the roof. Before 1800 this was the most common technique of coal mining.

A valuable early account of the pillar-and-stall method was written by a 'coal-viewer', or manager, in northeast England in 1708. From the base of the shaft, tunnels (known as headways) were driven no more than a yard and a quarter wide, and about 150 yards long. The coal was cut down by hewers, and then barrow men hauled the baskets of coal (known as corves) on sleds either to horse-drawn trams or to the base of the shaft itself where the corves were fixed to a rope. It was reckoned that it was feasible to bring out 160 corves a day if the pit was 120 yards deep. (By the end of the eighteenth century the Sheffield coal-viewer John Curr said that the capacity of a single corf was 19 pecks of coal, about 5½ cwt.) At the pit head the corves were unhooked by the banks men. They had an important role in implementing quality control, since the chief of the banks men was paid to ensure that the corves were properly filled with saleable coal and did not include slack.

A different technique, known as the longwall method, was pioneered in Shropshire in the seventeenth century. In this method the whole of the seam was removed, giving it a distinct advantage over the pillar-and-stall method. To prevent collapse, the space from which the coal had been removed (known as the 'gob' or the 'goaf') was packed with rock or slack, and the roof was supported by pit props.

Longwall mining subsequently spread to the East Midlands, Lancashire, Yorkshire and South Wales. In 1811 John Farey gave an account of the longwall method in use in Derbyshire, a valuable insight that also reveals how underground workmen combined their skills to work as a team. Colliers known as holers used a staff and pick to undermine the coal face to a depth of up to 30 inches, using wooden supports to prevent the seam collapsing. Then another set of colliers, known as hammermen or drivers, forced out the coal in large blocks. Next, a rambler broke the blocks down into manageable pieces and the loaders filled the coal into corves. Punchers

In this view of a pit head in Shropshire in the eighteenth century there is a horse whim for winding and a wooden waggonway for conveying the coal to a wharf on the River Severn.

and timberers then entered the pit to shore up the workings with wooden posts and underwood taken from plantations. Finally, men threw back the small coal into the gob.

The advantage of the longwall method was that it yielded more coal than the pillar and stall method. One disadvantage was that the mine was more prone to subsidence, but the major drawback was the combustibility of the slack coal used to fill the gob. Gob fires were sometimes serious enough to cause the closure of a pit – as happened at Oakthorpe Colliery (Leicestershire) in 1850 – but eventually encouraged sophisticated means of ventilating coal workings.

The horse whim or gin was a common method of raising coal from a pit, whereby a horse travels in circuits, winding a rope on a large horizontal drum. They were often installed by the sinkers and then taken over by the colliers when the pit went into production. They raised and lowered the corves, which contained variously coal and colliers entering and leaving the mine.

Some of the earliest innovations in the coal industry occurred at the pit head. Wooden railways, known as waggonways, were pioneered in the early seventeenth century

Haulage underground was greatly improved by laying rails, first of wood and only later of iron.

in Northumberland and Durham and in the East Shropshire coalfields. It was claimed that the use of wooden rails reduced the friction of wheels to the extent that one horse on a waggonway could pull as much as two horses and two oxen on a road. By the early eighteenth century waggonways with wooden rails could be found in most of the coalfields. They were used almost exclusively to carry coal from the pit head to the nearest water navigation, which in the northeast often meant to staithes where coal was tipped directly into coastal vessels. In places where waggonways were set at a considerable gradient an inclined plane was needed, by which loaded wagons descended by gravity and pulled the empty wagons back up at the same time. These were well established in Shropshire by the mid-eighteenth century, where wagons carrying 2 tons of coal were lowered from the mine entrance to river barges.

Coal was also hauled increasing distances underground, because to dig coal from ever increasing depths it was economical to dig a small number of shafts. In the early days this was achieved by baskets on sleds, pulled either by horses or humans. In the early eighteenth century wooden waggonways were introduced, creating an underground network more extensive than the waggonways above ground.

DEEP MINING

COAL MINING EXPANDED rapidly in the eighteenth and nineteenth centuries, thanks in large part to the expansion of Britain's iron industry. Coal, in its refined form of coke, was first successfully used for smelting iron ore to produce pig iron in the early eighteenth century; and before 1800 coal was the fuel for the new puddling process, the means by which pig iron was converted to malleable wrought iron. In this period the coal and iron industries combined: coalmasters invested in ironworks and ironmasters invested in coal mines. The success of the iron industry had a second beneficial effect. The invention and construction of steam locomotives, which was first sponsored by ironmasters, and the mass production of iron coupled with the development of rolling mills, made possible the railway. Railways transformed the coal industry, not only because coal was the fuel of steam locomotives. Railways could transport coal relatively quickly and cheaply to inland markets and to ports for export. As a consequence, coal became the principal domestic and industrial fuel in the nineteenth and much of the twentieth centuries.

The South Wales coalfield emerged as a major producer from the mid-eighteenth century, where iron companies opened coal pits to supply their furnaces. Then coal for export was developed, known locally as sale coal, following the railway boom that began in the 1830s. In 1855, at the peak of the iron industry, 8½ million tons of coal were dug in South Wales. In the next generation, development of sale

Opposite:
The chimney above the air shaft at Cwmbyrgwm Colliery in Torfaen is part of a landscape of pits and tips of this mid-nineteenth-century mine.

coal from the South Wales coalfield was so rapid that by the peak year of 1913, 620 collieries employing 232,800 people produced over 57 million tons of coal.

After the peace of 1918 there was a gradual decline in national output as demand at home was levelling off and coal for export became less economical. Most of the coalfields suffered, but especially those that had relied on the export market. South Wales, hitherto dominant in the British industry, suffered most. But there was growth in other areas, notably South Yorkshire and the East Midlands. New deep mines were sunk between 2,000 and 3,000 feet deep, to exploit thick and easily worked coal seams. Seven huge, state-of-the-art pits were sunk in the Dukeries area of Nottinghamshire in the decade following the First World War, including Clipstone, Harworth and Thoresby, all of which were designed to produce a million tons of coal annually. By 1930 nearly 20 per cent of British coal was mined in South Yorkshire and Nottinghamshire.

Almost all of the collieries in the United Kingdom came into state ownership under the National Coal Board in 1947. A significant number of small collieries, employing fewer

The Hetty Shaft of the Great Western Colliery, Trehafod, was sunk in 1875, and its engine house is the earliest of the surviving deep-mine engine houses in Wales.

than thirty men, were granted licences by the Coal Board that allowed them to remain in production. Small-scale collieries continue to operate in the coalfields after the large-scale deep mines have gone. Most of them are worked either from adits or drifts.

Nationalisation occurred at a time when the demand for coal had already peaked and the optimism of the early post-war years did not last. The steam engine was superseded by the internal combustion engine and for electricity generation coal faced competition from oil, gas, nuclear and, more recently, biomass. In the home, gas-fired central heating replaced the coal fire. The number of pit closures accelerated from the 1960s onwards, and continued after the industry was put back in the hands of private enterprise in the 1990s. As the getting of coal from deep underground became less viable economically, so opencast mining has come to prominence since the 1940s. Where it has occurred, notably at the head of the South Wales valleys, whole areas, including small settlements and archaeological evidence of early coal digging, have been dug away.

Big Pit, near Blaenavon, closed in 1980 after a working life of 120 years. It is the best-preserved colliery in Wales and is now its national coal museum.

Woodhorn Colliery in Northumberland opened in 1894 and closed in 1981, when over 450 men still worked underground. The surviving surface buildings now form the Woodhorn Museum.

In the era of deep mining collieries grew steadily in scale, winning coal from ever-greater depths. By the end of the nineteenth century some of the coal faces were over a mile from the shafts. Much of the increase in the scale of production was made possible by advances in coal-mine technology, the effects of which were visible at the pit head. The colliery in 1700 had perhaps a horse whim on the surface for winding, but little else. By the twentieth century the pit head was a complex of structures, including winding gear, engines for pumping and winding, ventilator fans and their engines, a screening area for grading coal, and coal washeries, joined later by pit-head baths, lamp rooms and canteens. As coal production became concentrated into a smaller number of mines, a colliery was a concern large enough to be served by its own workshops. On the surface was a network of railways and sidings to load the coal into trucks and remove it from site as soon as convenient.

The main technical problems of deep mining were threefold: the need to wind coal to the surface from great depths, to control the quality of air underground and to prevent groundwater inundating the workings.

Drainage had always been an issue with mining, but where workings were shallow, some reduction in groundwater could be achieved by digging surface ditches. A ditch was built across Newbold Moor in Leicestershire in 1554 for just this purpose. Drains were dug in mines to allow water to drain into a sump, but it was more difficult to remove water that did not soak away naturally. In the northeast of England in the early eighteenth century water was often raised in buckets with the same horse whim that drew up the coal. The contemporary travel writer Celia Fiennes noted a similar technique in North Wales in 1698. Waterwheels and wind mills were also tried. In 1610, miners at Stratton in Somerset built a waterwheel for mine pumping, a method that would enjoy considerable success throughout the Somerset coalfield. The last recorded example was working at Vobster Colliery until 1867. Elsewhere these simple measures were less successful. The anonymous author of *The Compleat Collier*, published in 1708, bemoaned the inadequacy of drainage methods as then practised, pointing out that several good collieries were rendered useless because

Crumlin Navigation pit, Caerphilly, was built in 1907 and closed in 1967. Finding alternative uses for colliery buildings has not always been easy.

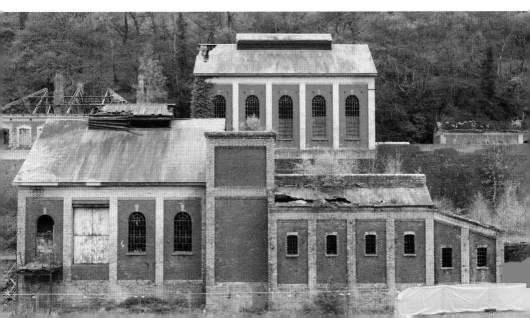

Murton Colliery, County Durham, painted in 1843 by John Carmichael, the year it entered full production. With three shafts it was one of the most ambitious mines in Britain.

the workings had flooded. The answer to these problems was the steam engine.

The steam engine devised by Thomas Newcomen was first employed in mine drainage in Staffordshire in 1712. In the next twenty years, until the patent expired in 1733, over one hundred of these engines were built at British collieries, but by the end of the century there were over a thousand of them in use. These simple beam engines lifted the water initially by suction from a sump at the base of the shaft, and then raised the water by a series of buckets. Newcomen's engines were simple but effective, in fact so simple that they could be taken down and re-erected on a new site if the pit for which they were built closed. Many of them remained in use well into the nineteenth century.

The second problem that beset collieries was ventilation, which became more serious as pits became deeper and workings more extensive. There were suffocating fumes, known as choke damp or black damp, and combustible gas, mainly methane, known as firedamp. Colliers learned to read the symptoms of firedamp by the colour of the flames in their candles. In shallow mines colliers resorted to wafting away patches of damp with ferns and gorse, but more effective methods were needed if coal production, and the depth at which coal was worked, were to increase. Ventilation was important not only in the extraction of poisonous or combustible gases, but in the supply of fresh air to mine workers, and for cooling the air underground, an important consideration as temperature was reckoned to rise by 1°C for every 40 metres' depth of working.

The pumping engine house is the only surviving building at the collieries of the British Ironworks at Abersychan, Torfaen, built in 1845.

Secondary shafts, known as upcast or air shafts, were employed to create a flow of air underground, which ensured that foul air did not stagnate. Where a mine had only a single shaft, ventilation could be achieved by dividing the shaft into two using wooden partitions known as brattices. To improve the draught, a fire would be lit in a bucket suspended in the shaft; later, a small furnace was built at the base of the shaft. Management of air flow was aided by the use of trap doors, which meant that air could be directed into current workings and diverted when necessary to clear out gases from old workings. By 1816 it was calculated that a mine extending 500 yards square underground would require the air to travel along passages for 18 miles.

Burning was one of the most common forms of miner's injury, although in only a minority of cases was an explosion severe enough to receive widespread publicity – for example in 1705 when thirty colliers were killed in a blast near Newcastle-upon-Tyne. Since the purity of the air could not be guaranteed, a safety lamp was needed to replace candles. A series of fatal accidents, notably the explosion at Brandling Main Colliery on Tyneside in 1812, which killed ninety-two miners, brought the problem into sharp focus. The Society for the Prevention of Accidents in Coal Mines asked Humphry Davy, the scientist employed by the Royal Institution, to visit the northeast and to devise a safety lamp. Davy's wire-gauze lamp was based on the discovery that a flame could not pass through the very small apertures of the gauze and could only ignite the gases inside it. The earliest version was trialled successfully at Hebburn Colliery in 1816.

Take up of lamps by Davy, and a rival one by George Stephenson, was quickest in the northeast, but in general naked lights were still used in Britain's coalfields for several decades, not least because they were brighter than safety lamps. Speaking to the Children's Employment Commission in 1842, the agent of the Penydarren collieries in Merthyr Tydfil explained that 'the miners have Davy lamps to go into the workings in the morning to see that the air is pure, but they never work by them – they are never allowed to work in places where the air requires them to use Davy lamps'. Despite these assurances, underground explosions were common in nineteenth- and early twentieth-century coal mining, and with them came an appalling death toll. Responsibility and blame was contested.

The miner's safety lamp was invented in the early nineteenth century but only slowly did it become adopted across the entire industry.

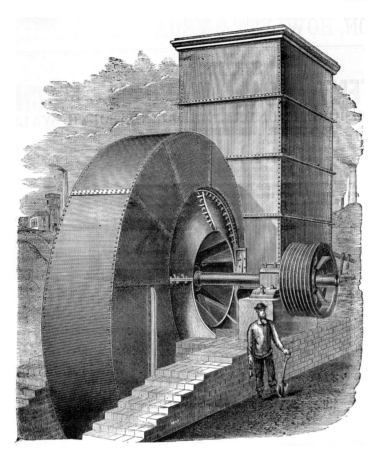

This advert from the *Colliery Guardian* in 1883 is for a Schiele-type fan, which is enclosed on the left, drawing out foul air through the broad chimney, or *évasée*.

Inquests often put the cause down to the careless practices of (invariably deceased) miners, although explosions often occurred where men were working with candles in parts of the mine that were supposed to be free of firedamp.

The clear benefits of the safety lamp quickly became apparent to coal owners, however, because they are excellent firedamp detectors. It meant that coal could be worked in underground areas hitherto considered too dangerous and was one of the factors that allowed deeper and more extensive seams to be exploited. In the northeast there were more fatalities in the eighteen years after the introduction of safety lamps (538) than in the eighteen years before they were

North Duffryn was a small early-nineteenth-century colliery near Merthyr Tydfil. Coal was raised in a large iron bucket by the engine on the left. On the right is a later Waddle-type ventilator fan.

introduced (447), largely due to the expansion of operations that the safety lamp had made possible.

Furnace shafts remained in use for mine ventilation for much of the nineteenth century. They were replaced by the much more efficient fans that drew air up the ventilation shafts and discharged the foul underground air into the open. The most common types were named after the engineers who invented them – the Guibal, Schiele and Waddle fans. Air was drawn out of a broad squat chimney known as an *évasée*, betraying the continental origin of the technology. Fan ventilators were efficient enough to allow deeper mine working, but they were not a failsafe, and the consequences of igniting gases in large-scale mines employing hundreds of people were catastrophic. Some of the underground explosions were so powerful that pit-head structures were destroyed. The disaster at Oaks Colliery near Barnsley in 1866 began with an explosion that killed 334 miners. Rescue attempts were initially hampered because the cages in the shafts had been destroyed, and were thwarted when a second explosion the following day claimed a further twenty-seven lives. Subsequent explosions occurred on subsequent days and a decision was taken to stop up the shafts to prevent further

The only surviving water-balance head gear has been re-erected at Big Pit, Blaenavon. The pulley is set above two shafts, used alternately for raising coal and lowering the water-filled tank.

fires, even though not all of the bodies had been recovered. The pit was re-opened only when new shafts were sunk.

Deep pits required powerful means of raising coal from underground. In the South Wales coalfield most of the early workings were relatively shallow. This meant that, instead

This view of a nineteenth-century Staffordshire colliery shows the simple beam engine put to winding duties in a divided shaft.

The engines in the power hall at Astley Green, Lancashire, were manufactured in 1912 by the engineers Yates and Thom. The winding drum is on the right.

Since 1911 all pit-head frames have been built of steel or concrete, but no two are exactly alike. This example from Penallta, in the Rhymney valley in South Wales, was built in 1906.

of using horse whims or steam engines to raise coal, they could use water-balance lifts, whereby a tank of water acted as a counterweight to the coal at the base of the shaft. Self-evidently such mechanisms could only work in shallow pits where there were no drainage problems, but they were common in South Wales up to the mid-nineteenth century.

Steam-engine technology advanced rapidly in the nineteenth century, and the machines were soon adapted to winding duties. One consequence of this was the characteristic head frame. The head frame and its sheaves (or pulleys) have become the enduring image of the coal industry, and yet they were conceived as only temporary structures. Early examples were built of pitch-pine and had a working life of about thirty years. Wrought-iron frames were introduced in the mid-nineteenth century, to be superseded by steel

Corves were hooked individually to the rope at the base of the shaft for winding to the surface.

and later on by reinforced concrete, but for safety reasons the Coal Mines Act of 1911 stipulated that new head frames could no longer be of timber. Shaft capacity was improved by the introduction of cages steadied by guide rods in the 1840s. At about the same time corves were given up in favour of wheeled wagons (known as drams, tubs, dans or hutches)

The efficiency of winding was greatly improved when the tubs could be wheeled directly into cages for winding to the surface.

Above: This two-tier cage at Astley Green, Lancashire, was typical of the lifting arrangements at collieries.

Above right: Tubs were much more efficient than corves as they were easy to manoeuvre underground and on the surface.

which, together with the increasing power of steam engines, allowed greater amounts of coal to be lifted for the same amount of time.

Innovations also occurred underground. Underground haulage was improved by engines powered by steam, and later by compressed air and electricity, which made haulage cheaper and less labour intensive, although there was still plenty of work for the humble pit pony. Output was increased by the use of explosives, which began in the early nineteenth century and were used to bring down or just loosen the coal. Apart from the safety implications, one disadvantage of explosives was that they increased the proportion of slack coal and brought up more waste (or dirt) than coal cut by hand. The size of slag tips therefore increased with massive coal production from larger pits, making a significant impact on the coalfield landscapes.

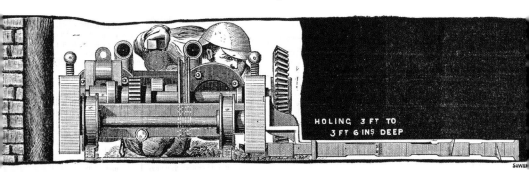

HOLING 3 FT TO 3 FT 6 INS DEEP

Modern coal-cutting machines, like this one at Kellingley, Yorkshire, cut the seam and loaded the coal onto conveyors.

Powered drills first appeared in the late nineteenth century, but one of the problems with them was the increase in the dust thrown up and breathed by the miners.

Until the early 1900s coal was still dug mostly using the longwall method developed in the seventeenth century. Mechanical coal cutters, mainly powered by compressed air, and pneumatic picks found favour in some coalfields. Conveyors were introduced that took the coal from the face to be loaded into tubs. The problem with the early mechanical devices, however, was that only a small proportion of time could be allotted to cutting the coal. Separate shifts were required to load the coal and take it away from the face, while a third shift positioned the machinery for the next cutting shift. The most successful of the early machines that could cut the coal and load it on to conveyors was the Meco-Moore Cutter-Loader manufactured from the 1930s. It was cumbersome in operation, however, and required about fifteen men to operate and manoeuvre.

By 1945 about 72 per cent of coal was cut mechanically. Mechanisation increased after the war, with renewed efforts

Opposite: This early coal-cutting machine shows how the seam was holed-out in a traditional manner.

to improve efficiency by designing a machine that would allow a greater proportion of time cutting coal. The great innovation of the post-war era was the Anderton shearer loader, invented by the general manager of the St Helens area of the NCB and developed by the Coal Board itself. By 1966 half of coal cut in British collieries was by this machine. A more complicated type of cutter, known as a trepanner, also found favour because it could be used to work thinner seams and cut the coal into larger pieces, which made it useful in cutting coal for the domestic market. The Anderton shearer loader, and variants of it, produces much more small coal and dust, which was a disadvantage at first, but when the market for domestic coal decreased and for small coal increased, the Anderton came into its own. Other improvements were made underground after 1945. Hitherto the roofs of the pits had been held up by pit props, but these were superseded by hydraulic props.

When it was brought to the surface the coal was screened, or sieved. Coal was tipped on to picking belts, moving

The pit-head layout at Hebburn Colliery, near Newcastle-upon-Tyne, in the mid-nineteenth century included the engine on the left. To the right of the wooden head gear is the heapstead where the coal was screened.

conveyors from which the largest lumps of coal and any dirt were removed. Coal was then tipped into screens – essentially large sieves with iron grates set at about half an inch apart – to remove the small coal. Screens were developed over the nineteenth century to allow finer grading of coal to suit different markets. Screening plants were usually raised on stilts to allow for the coal to be fed through chutes into railway wagons beneath them, a structure known as the heapstead. Coal washers were also used from the mid-nineteenth century, and developed further in the twentieth. Clean coal is light and floats in a tank of water, where it was screened off. Coal with dirt in it sinks so, when it was removed, it was crushed to separate dirt from the coal before being washed again in a separate tank.

The twentieth-century heapstead by the Hesketh shaft at Chatterley Whitfield is a rare survival and is where the coal was screened before being loaded onto railway wagons.

GOING UNDERGROUND

COAL MINING WAS always a labour-intensive industry, demanding people with different levels of skill, stamina and physical strength. A skilled coal-face worker was a valuable asset if he could cut the coal efficiently and with little slack. Borers and sinkers too were men with special skills. In the coal trade these skills tended to be passed down to the next generation, who had started in the pit as boys engaged in unskilled labour. Rapid expansion of the industry, however, brought new blood, ensuring that there were always first-generation miners in the pits up until the industry peaked in 1913.

Over time, and as the technology of getting coal advanced, coal mines became increasingly sophisticated operations. A valuable snapshot is provided by a list of people engaged in 1812 at Temple Main Colliery on the Tyne. It employed 440 men and boys, 128 of whom worked on the surface. There were 106 'hewers' cutting the coal and forty-two 'putters' hauling it away, but there were also tramway layers and horse keepers working underground. Apart from maintenance staff such as wrights, joiners and engine keepers, the large contingent of surface workers included waggonmen and thirty-seven screeners and wailers (the latter usually boys) who picked out the dirt from the coal brought to the surface.

In the eighteenth and nineteenth centuries the age profile of the workforce was predominantly young – unsurprisingly given the dangerous and arduous nature of the work. Of

Opposite:
The collier, as published in George Walker's *Costumes of Yorkshire* in 1814.

Picking out dirt from a conveyor at the pit head was monotonous work carried out in all weathers and temperatures.

the ninety-two victims of the Felling Colliery disaster on Tyneside in 1802, forty-five were under twenty years of age (although two of them, John Pearson and Isaac Greener, were sixty-four and sixty-five respectively). Many of the tasks at the colliery were classed as unskilled light work, for which women, girls and young boys could be employed, at least until 1843 when children under ten and females were forbidden from working underground. Elizabeth Pendry and Annie Tonks, aged six and twelve respectively, were among those who perished in a pit explosion at Cwmgwrach, near Neath, in 1820. Matthias Dunn, a colliery manager from the northeast, was smitten after encountering an onsetter (the person responsible for hooking the coal container to the rope at the bottom of the shaft) on his visit to Lancashire in 1825. She 'sported a pair of golden ear rings', and wore 'a shift of flannel, a pair of huge white flannel trousers, a short bedgown and overall'. He went on to describe the putters working at a seam only 4 feet high, who were girls dressed in trousers and jackets.

Coal mining was a way of life that started in childhood. The Children's Employment Commission of 1842 documented the working lives of these young people, although only in the coalfields of South Wales, Yorkshire, Lancashire and eastern

Women were employed in unskilled work, but it could be physically arduous, as shown in this view of coal workers in Scotland.

Scotland were girls employed underground by the middle of the nineteenth century. Some of the inspectors expressed their horror at the unseemly nature of the work: 'One of the most disgusting sights I have ever seen was that of young females dressed like boys in trousers crawling on all fours with belts around their waists and chains between their legs.' From the age of seven upwards, boys undertook similar jobs in hauling but were also engaged in driving horses underground and in operating the air doors that controlled ventilation.

A girl putter in 1842 wears a belt around her waist with an attached chain that passes between her legs to pull a tub of coal in a confined space.

After women were banned from working underground in 1843 many found work on the surface. The pit-brow women of Lancashire worked at the pit head until the 1950s.

PIT BROW GIRLS 6.

Although many children were taken underground to work for their fathers, there were too many young unmarried hewers and orphaned children for this always to have been the case. Once they were forbidden from working underground women often found employment at the pit head, although the nature and content of their work varied between the coalfields. They remained in the workforce longest in the Lancashire coalfield, where they were known as 'pit-brow lasses'.

The underground world was a world apart, but it was open to gentleman travellers and journalists if they were brave enough. When he was invited down a local mine in the 1850s the Shropshire journalist John Randall noticed that 'a group of pit girls share a sly joke at your expense'. Randall was struck by 'the curious indescribable smell' of the underground stables and the discomfort of working underground: 'the atmosphere is oppressive and you perspire freely'.

One of the important jobs for children in the mines was operating the air doors, which was a lonely and long day in the pitch black for a small child.

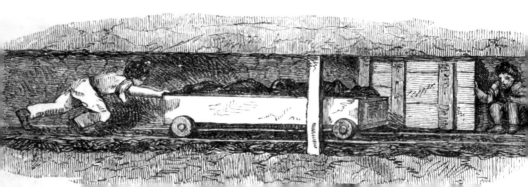

Two hewers working underground in the early twentieth century.

With genuine trepidation, in 1853 a Congregationalist minister, John Leifchild, entered the tub, effectively just a large iron bucket, swinging at the top of a 1,680-foot shaft at a Sunderland pit, and then descended what acted as the upcast shaft, in which the rising fumes and smoke made his fellow passengers crouch low in the tub to breathe more easily. At the base of the shaft a piece of clay was pushed between his fingers in which a candle was inserted. He noted the little boys employed to open the air doors, and the loneliness of

their long shifts without lights. The height of the workings depended upon the thickness of the seam, and when the seam was very thin the men worked with their picks lying on their sides. Leifchild encountered a seam only 3 feet 6 inches high. The hewers were 'the hardest

A coal miner crawls through a confined space with his safety lamp, showing the conditions that miners often worked in, even into the twentieth century.

Working in cramped conditions, a hewer holes out the base of the seam at Denby Colliery in Derbyshire, in 1898.

labourers in the pit' and the operation of the mine was organised around their needs, but the putters' jobs were only a little less exhausting, pushing the tubs on the rails, in teams when the incline was steep. Leifchild characterised them as fit lads who drank copious amounts of water to replace the losses through perspiration, and who ate as much as they could. He watched them 'devour huge hunches of bread and cheese, with a bone or so of meat, drinking at that time cold coffee or milk out of tin canteens'.

Miners hole out the base of a seam in preparation for blasting.

ACCIDENTS UNDERGROUND

In this cramped and claustrophobic world accidents were common, and one Shropshire coalmaster reckoned that about 10 per cent were fatal. Rock falls were often fatal to underground workmen and were probably the most common form of accident, although it was multiple fatalities that attracted notice outside the coal community. Winding procedures were notoriously unreliable. Corves could swing out of control in shafts and winding engines did not always stop when their load had reached the surface, sending the corves and the people hitching a lift in them over the tops of the pulleys. In 1821 eleven people were ascending a shaft at Norcroft Colliery in Yorkshire when the chain broke, plummeting the miners to the base of the shaft; nine of them lost their lives. Twelve men died in similar circumstances at Wellsway Pit in Somerset in 1839, when the winding rope broke and the men plunged 750 feet from the top to the bottom of the shaft. Ground water was also hazardous. Thirty-two children, between the ages of eight and seventeen, were drowned at Huskar Colliery near Silkstone in Yorkshire in 1836, after the pumping engine was rendered inactive and a drift mine flooded. In 1837 the roofs collapsed and the sea broke in to three pits on the coast at Workington, killing twenty-seven colliers and twenty-eight horses, but it occurred when the shifts were changing and so there were comparatively few workmen underground at the time.

A newspaper report of an underground explosion in Scotland in 1816 conveys the horror of the experience and the difficulty of mounting a rescue operation. After the initial explosion 'enveloping the unhappy miners in quick burning fire' there was a 'momentous silence', followed by an inrush of air so powerful that 'with the noise of the loudest thunder [it] sweeps before it into horrible ruin and destruction the

Methods of descending shafts by clinging to the rope seem scarcely credible now, but such practices were common until enclosed cages were introduced in the mid-nineteenth century.

Wives and mothers gather at Hartley Pit in Northumberland in 1862 after news spread of an explosion. The disaster claimed 204 lives.

unhappy miners, with the horses, carriages and working implements, and dashes, mangles and buries them in one common ruin amid the rubbish and timbers'. Any survivors of this phase would then become enveloped by foul air, known as 'after damp' and containing a lethal cocktail of gases, including carbon monoxide.

Early rescue teams had some notable successes, however. At Tynewydd Colliery, in the Rhondda valley, in 1877 fourteen miners were trapped by an inrush of water from a neighbouring abandoned colliery. Five drowned, and four were soon rescued. The remaining five were trapped underground for ten days in an air pocket before rescuers tunnelled through to reach them, in a mine where accumulated gases made a dangerous atmosphere for both the trapped and the rescue team.

Not until 1902 was a regional mines rescue station opened, at Tankersley in South Yorkshire, although provision of mines rescue services was made compulsory after the 1911 Coal

This engraving, published in *Le Monde*, depicts injured men being hauled out of Ferndale pit in the Rhondda valley in 1867, a disaster that claimed 178 lives.

Mines Act. Shortly afterwards they acquired their best-known piece of kit. Canaries were introduced into coal mines because they are more susceptible to odourless but harmful gases than humans. A sign that the bird was in distress was evidence that the air was impure. Canaries delivered over seventy years' service before they were phased out in favour of more prosaic gas detectors in 1986.

PAY AND CONDITIONS

Coal mining is a dirty job and miners washed themselves in the bathtub at home. The Coal Mines Act of 1911 obliged colliery owners to provide baths, but only if a two-thirds majority of miners at a pit requested them and contributed to the cost of building them. As a result, in 1920 only ten collieries provided bathing facilities. Legislation passed in 1926 allowed the Miners' Welfare Committee to reap a royalty on coal production that financed the construction of the baths, but it did not make significant progress until the 1930s. Bath houses contained multiple shower cubicles, clean and dirty locker rooms, an ambulance room, and often facilities for mundane chores like boot greasing and bottle filling.

Two mine rescue workers are pictured at West Stanley pit, County Durham, in 1909, where 168 lives were lost in the wake of an explosion, 120 by carbon monoxide poisoning.

Pit-head baths were a major symbol of progress in the coal industry. Other innovations that began in the inter-war years contributed to improved underground safety. Safety helmets were introduced in the 1930s, to which battery-operated safety lamps could be fixed. Manriding cars delivered miners the long distances from shaft to coal face quickly and efficiently, although pit ponies still hauled coal underground. The hazards of firedamp were reduced by drilling methane drainage holes into the coal face and attaching pipes that

Ponies, like this one at Brinsley Colliery, Nottinghamshire, in 1913, continued to work in pits in the twentieth century. Improvements in their welfare were covered by the 1911 Coal Mines Act.

could pump it to the surface out of harm's way. Dust, however, proved a more intractable problem. Coal-cutting machines increased the amount of dust in the pit, the long-term consequence of which was a significant rise in the incidence of pneumoconiosis, a condition that is slow to appear but seriously affects the long-term health of sufferers.

Before nationalisation there were no standard rates of pay on a national scale. Skilled miners were comparatively well paid, reflecting the nature of the job, but only if they remained fit and healthy. The age profile of the mining workforce was biased towards youth, and there were many jobs classed as unskilled or light work that commanded much smaller wages. Wage levels were further complicated by the need to compensate for occasional stoppages, allowances of beer and money for candles, gunpowder and lamp oil. In most places, however, piece rates were paid to hewers, measured in corves. For example, at Harberlands Colliery in Derbyshire in 1795 hewers were paid 18d per dozen corves, and 12d for the equivalent of small coal. The problem with this method was the place where the baskets were measured. Miners objected if the coal was not weighed at the pit head, because the further it was taken from the mine the more of it would be pilfered and the miners would lose out. For most of the other trades a piece rate was impractical and day rates were paid instead. In areas such as Shropshire the charter-master system was operated. The charter-master was a middle man between mine owner and miner, who organised the underground labour and paid the workmen on agreed terms.

The length of shifts also varied considerably, but was usually between eight and twelve hours per day. With a twelve-hour day there were breaks of up to one-and-a-half hours. In terms of shift length, miners therefore worked less than some other industrial workers, but their labour was strenuous and carried out in a difficult physical environment.

Employers introduced subscriptions to pay for surgeons to attend to work-related injuries. In many instances mine owners also accepted some responsibility for the dependants of those killed in colliery accidents. There were plenty of them. A survey undertaken in the second decade of the nineteenth century identified 464 colliery widows on Tyneside alone. Ex gratia payments were made to some of them, including one-off compensation payments and payment of burial fees, but they were at the discretion of individuals. Although none of the payments were exactly generous, widows were not generally told to quit their house and were usually given an allowance of coal. But the generosity of coal owners was not always sufficient. When the Workington Coal Miners' Society was established in 1793, its primary stated objective was to make provision for those who had suffered accidents, and for their dependants. It was part of an emerging movement. Friendly Societies looked after the long-term welfare of mining families from weekly subscriptions – the spirit of self help that sustained communities before welfare was nationalised in the twentieth century.

UNIONISATION

The means by which miners improved their lot was collective action. From the eighteenth century miners pooled their resources to act collectively with employers and to organise their welfare. The early combinations were often ephemeral and local, and operated in the face of considerable hostility in the period of the Anti-Combination Laws of 1799–1824, but they evolved into regional trade unions in the nineteenth century. Autonomous regional unions reflected the regional nature of the coal industry, but gradually they pooled their efforts, forming the Miners' Federation of Great Britain in 1888, to which regional unions affiliated themselves. The most powerful and most militant by that time was the South Wales Miners' Federation (the Fed). Unions were active in campaigning on wages, working hours and health and safety. Improvements in underground working conditions, and in raising the age to which children should be permitted to work underground, were not won without hard-fought campaigns. As the coal industry became ever more prominent in the nation's economy, so the effects of industrial action were widely felt. The strike in South Wales in 1911 and the General Strike of 1926, in which the coal industry was prominent, were both fought against the background of wage cuts and, latterly, of job cuts. Miners in Senghenydd, scene in 1913 of Britain's worst colliery disaster, were given only a day's notice of the closure of the pit in 1928. In 1945 the Miners' Federation of Great Britain was reorganised to form the National Union of Mineworkers, which remained a powerful force in the industry until the end of the 1984–5 miners' strike (a protracted struggle to prevent the mass closure of state-owned collieries) and spans the era of nationalisation.

THE PIT VILLAGE

THE INFLUENCE OF the coal-mine owner extended beyond the coal pit to include the pit village. Now that the collieries and the tips that once dominated the landscape have largely gone, the pit village is usually the most direct reminder of the industry that was. Pit and pit village were part of an interlinked living and working community. Miners received payments in kind, in the form of subsidised housing and an allowance of coal (not exceeding domestic needs, in order to prevent miners from selling the surplus), which was sometimes free, or available below the market price.

Since the location of mines was determined by geology rather than by existing infrastructure, mine owners usually needed to provide housing for their workmen, which amounted to a substantial investment in any individual mine. The opening of a coal mine rapidly swelled the population of rural parishes. In 1801 the parish of Hetton in County Durham had 253 inhabitants, but thirty years later the growth of mining villages had swelled the population to nearly 6,000. Development of coalfield settlements had accelerated by the end of the nineteenth century. Over 113,000 people lived in the Rhondda valley in 1901, compared with fewer than a thousand half a century earlier.

Mining settlements of the early nineteenth century were of short rows of terraced houses. The standard was at least as good as that of the homes of agricultural labourers; nor were they always as overcrowded as is often supposed. A survey

Opposite:
The grand extension to the Park & Dare Workmen's Institute, Treorchy, was built in 1913. Funded by miners' subscriptions, it is a monument to the aspirations of working men.

Next page:
Cwmparc in the Rhondda valley is a typical colliery settlement of late-nineteenth-century South Wales.

Rural isolation of coal pits meant that coal masters needed to build houses for the workforce. These houses at Waldridge Colliery, near Chester-le-Street, stood right next to the colliery.

Right: The pit village at Brithdir, in the Rhymney valley of South Wales, was begun in the last decade of the nineteenth century. The pit it served is long gone.

in 1757 of Hartley Colliery in Northumberland recorded an average of just over four people per house in a village of sixty-seven households. The Durham miner George Parkinson wrote affectionately of his childhood home in the 1830s, and his description reminds us that coal mining was a rural industry. His row of brick houses faced 'a meadow through which ran a clear burn', but the houses themselves were spartan: 'Behind the door a ladder led to the upper room or loft close to the tiles, which were not hidden by any plaster or wooden ceiling. The flooring boards of the loft were laid loose upon the joists.' Even when running water was supplied to houses there was no bathroom; the tin bath, hanging on a nail, was characteristic of the miner's home well into the twentieth century.

However, the rapid expansion of mining and the building of large settlements of closely spaced houses created problems in the nineteenth century. Robert Haddow's Scots colliery village, described in 1888, was towards the meaner end of the scale: 'The colliery village is, as a rule, the very embodiment of dirtiness, dreariness, and rough squalor. Long rows of dismal brick houses face each other, and standing between them are

the sanitary offices, which in hardly any one case are properly kept in order.' Older housing stock in the industry was made worse by the fact that miners rented them for very little, or had them rent free (especially in northeast England). Landlords therefore had little incentive to fix the damp or the leaking roof and, in any case, a pit village

Alfred Pollard, a miner from Eastwood, Nottinghamshire, washes himself at home in 1913. A tin bath was in every miner's home.

lasted only as long as the pit, a further disincentive to long-term improvements. In the 1850s John Leifchild bemoaned the state of pit villages in the northeast for their lack of space and simple sanitation, but many of the same houses survived long enough to be included in a survey of the coal industry undertaken in the 1920s. At any given time the best miners' housing was invariably the most recently built.

In most coalfields established by the early nineteenth century there were houses that had a single room downstairs and another in the loft reached by a ladder in which families slept. Larger houses, two-up, two-down, offered more comfort and privacy, although the rooms were small, as little as 10 feet square. There were also communal facilities. The *Glasgow Herald*, reporting in 1875, described Rosehall near Glasgow as having a closet for every three houses and a wash house for every six. These types of arrangements, along with communal bake houses, were common, even if they were often inadequate. Standards of accommodation improved in the latter part of the nineteenth century. In South Wales, for example, houses in the valleys were superior to those found in other parts of the coalfield, like Merthyr Tydfil, which had developed a century earlier with the rise of the iron industry.

Woodlands, shown here under construction in 1907, was built for the Brodsworth Main Colliery in South Yorkshire, and was a new type of pit village, with a rural English character and ample open space.

Not all housing was owned by collieries, however, and in the twentieth century mining companies decreased their ownership of houses. Building societies and clubs provided funds for houses and local authorities built significant quantities of new homes. In those coalfields experiencing rapid growth, like Nottinghamshire and Kent, coal owners were still prepared to invest in new housing schemes, although much of the investment was channelled through the Industrial Housing Association, which in the 1920s built 12,000 new houses in the East Midlands and South Yorkshire, all of high quality. On a visit to the Dukeries of Nottinghamshire in 1927 the *New Statesman* praised the new pit villages: 'the narrow meanness of the owners has been discarded. They have shown a wider outlook – a realisation of the social importance of cleanliness, of the social value of comfort.'

Outsiders did not always look kindly on coal miners. Mining was a dirty occupation and, with no facilities for cleaning themselves before leaving work, the appearance of men with blackened faces and clothes exacerbated the negative perception of them. Mine owners disdained their habits and sometimes even criticised them for being dirty. Sir John Clerk of Penicuik, Midlothian, despaired of his workmen at Loanhead Colliery, for their 'profaneness and immorality, particularly excessive drinking … fighting … cursing and swearing, or taking the Lord's name in vain'. But their supposed hard-drinking was of men engaged in dangerous, physically demanding and thirsty work. Even so, this is only a partial view of miners' communities and not a representative one. The standards of coal-mining communities were no different from those of the rest of society. In fact there are many other contemporary accounts that emphasise the cleanliness of the collier's family home. In 1833, for example, the Factories Inquiry Commission found the homes of colliers in Glasgow 'usefully and cleanly furnished' and that 'the collier women were very clean in their appearance'.

The problem of rapid growth in the industry was that pit villages were dependent upon one industry only. In the Rhondda valley, for example, some three-quarters of adult males were employed at the pits. During the slump of the 1920s thousands of people left Wales and the northeast to settle in southeast England where they could find work – over a quarter of Rhondda's population left in the 1920s. The coal

In 1951 people gather at Easington Colliery on hearing of a pit explosion. Major accidents had a profound impact on communities dominated by this one industry.

The Rhondda
Baptist Chapel,
with its attached
Sunday School,
epitomises
the religious
and cultural
life of mining
communities in
Wales.

industry had a surfeit of employees in almost all periods, except for the major wars; from December 1943 a small number of conscripts were chosen by ballot to work in the mines.

The growth of large pit villages from the final quarter of the nineteenth century created communities in which almost everybody was working class. It bred a strong collective culture where personal ambition found its way to the union lodge or the pulpit. Britain's coalfields became strongholds of religious nonconformity and the temperance movement in the nineteenth century. The growth of the coal industry in the final quarter of the eighteenth century coincided with the rise of Methodism. Pit villages were captive audiences for preachers, where many people formulated their rules of right living and where, by joining forces to establish a chapel, the working classes could assert their independence. Membership of a chapel community became a means to a cultural life. Early chapels were small, sometimes just converted cottages. The more ambitious chapels still to be found in the coalfields are nearly always second- or third-generation chapels that testify to the success of the nonconformist denominations in establishing and growing their congregations.

The Hetton Silver Band, here showing off one of the trophies they won in local championships, were regular performers at the Durham Miners' Gala.

Slowly the aspirations of coal miners, as with other industrial workers, changed. Literacy improved as church and chapel attendance increased. Schools, Friendly Societies, savings banks and trade unions gave miners a stake in the future. Nonconformists were active in opening schools and stimulating cultural life, the best-known manifestations of which were the brass band and male-voice choir. In South Wales and elsewhere, workmen's institutes became prominent, voluntarily funded out of miners' pay, which provided libraries, reading rooms and gymnasia. In Mountain Ash in

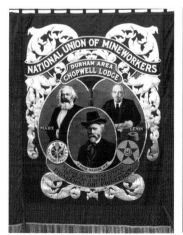
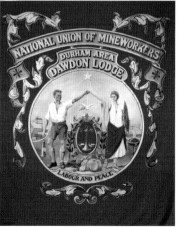

Union banners have long been carried in procession at the Durham Miners' Gala, where individual union lodges come together to celebrate the coal-mining community and its heritage.

Two young boys lead the Easington lodge banner and brass band at the Durham Miners' Gala in 1950.

the Cynon valley the institute erected by workmen of the Nixon's Navigation Collieries also included a theatre, billiards room and swimming pool. It bred a culture of ambition and self-improvement that nurtured many talents, in people like Aneurin Bevan, founding minister of the National Health Service, who had come a long way from his first job at Tytryst Colliery near Tredegar.

The Durham Miners' Association was established in 1869 and two years later held its first gala. It grew into an annual event, known as the Big Meeting, at which miners marched behind brass bands and their union banners, many of which portrayed their political heroes. There are no deep mines left in Durham, but the gala survives as a celebration of the coalfield community. To the tune of the brass bands banners are still carried in procession past the County Hotel at Old Elvet, Durham, before everybody converges on Durham racecourse where rallying speeches are made. It is the best kind of tradition, sustained by people who know their own history.

PLACES TO VISIT

Information about the numerous colliery remains in Britain can be found by searching online at www.heritagegateway.org.uk for England, www.coflein.gov.uk for Wales, and canmore.org.uk for Scotland.

   Several former collieries have been opened to the public and several museums have significant displays on the coal industry. Visitors are advised to check opening times.

Astley Green Colliery Museum, Higher Green Lane, Tyldesley, M29 7JB. Telephone: 01942 708969. Website: www.agcm.org.uk

Beamish Museum, Beamish, County Durham, DH9 0RG. Telephone: 0191 370 4000. Website: www.beamish.org.uk

Astley Green Colliery, the best preserved of the former Lancashire collieries, retains a magnificent power hall, the heartbeat of the colliery operations.

Bestwood Country Park, Park Road, Bestwood, Nottinghamshire, NG6 8TQ. (The former Bestwood Colliery, No 2 Pit, winding house and headgear.)

Big Pit National Coal Museum, Blaenavon, NP4 9XP. Telephone: 0300 1112333. Website: www.museumwales. ac.uk/bigpit

Black Country Living Museum, Tipton Road, Dudley, DY1 4SQ. Telephone: 0121 557 9643. Website: www.bclm. co.uk

Cefn Coed Colliery Museum, Neath Road, Creunant, SA10 8SN. Telephone: 01639 750556. Website: www.npt.gov. uk (Open May–September.)

Chatterley Whitfield Colliery, Stoke on Trent. Website: www.chatterleywhitfieldfriends.org.uk (At the time of writing only open on Heritage Open Days, or by prior arrangement.)

Elsecar Heritage Centre, Wath Road, Elsecar, Barnsley, S74 8HJ. Telephone: 01226 740203. Website: www.elsecar-heritage.com (Includes the only surviving atmospheric beam engine.)

Haig Colliery Mining Museum, Solway Road, Kells, Whitehaven, CA28 9BG. Telephone: 01946 599949.

National Coal Mining Museum for England, Caphouse Colliery, New Road, Overton, Wakefield, WF4 4RH. Telephone: 01924 848806. Website: www.ncm. org.uk

National Mining Museum Scotland, Lady Victoria Colliery, Newtongrange, EH22 4QN. Telephone: 0131 663 7519. Website: www.nationalminingmuseum.com

Prestongrange Industrial Heritage Museum, Morrisons Haven, Prestonpans, EH32 9RX. Telephone: 0131 653 2904. Website: www.prestongrange.org

Rhondda Heritage Park, Lewis Merthyr Colliery, Trehafod CF37 2NP. Telephone: 01443 682036. Website: www. rhonddaheritagepark.com

Caphouse Colliery in South Yorkshire closed in 1985 but reopened as the National Coal Mining Museum for England.

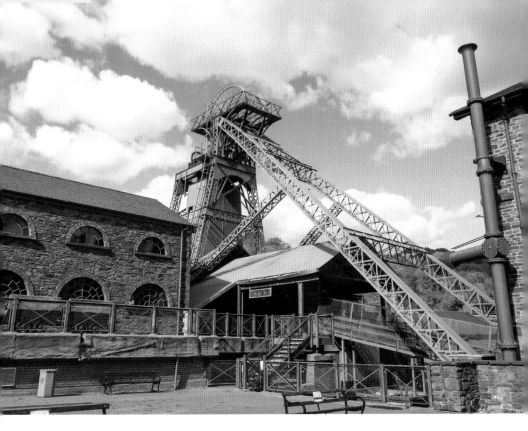

Lewis Merthyr Colliery, Trehafod, closed in 1983 and is now preserved as the Rhondda Heritage Park.

South Wales Miners Museum, Afan Forest Park, Cynonville, Port Talbot, SA13 3HG. Telephone: 01639 851833. Website: www.south-wales-miners-museum.co.uk

Summerlee Museum of Scottish Industrial Life, Heritage Way, Coatbridge, ML5 1QD. Telephone: 01236 638460. Website: http://culturenl.co.uk/summerlee (Contains a reconstructed mine and miners' cottages.)

Washington F Pit Museum, Albany Way, Washington, NE37 1BJ. Telephone: 0191 553 2323

Woodhorn Museum, Queen Elizabeth II Country Park, Ashington, Northumberland, NE63 9YF. Telephone: 01670 624455. Website: www.experiencewoodhorn.com

FURTHER READING

BOOKS

Ashworth, William. *The History of the British Coal Industry,
V: 1946–1982*. Clarendon Press, 1986.

Ayris, Ian and Gould, Shane. *Colliery Landscapes: An Aerial
Survey of the Deep-mined Coal Industry in England*.
English Heritage, 1994.

Church, Roy. *The History of the British Coal Industry, III:
1830–1913*. Clarendon Press, 1986.

Flinn, Michael. *The History of the British Coal Industry, II:
1700–1830*. Clarendon Press, 1985.

Hatcher, John. *The History of the British Coal Industry, I:
Before 1700*. Clarendon Press, 1993.

Hill, Alan. *Coal: A Chronology for Britain*. Northern Mines
Research Society, 2002.

Royal Commission on the Ancient and Historical
Monuments of Wales. *Collieries of Wales: Engineering and
Architecture*. RCAHM Wales, 1995.

Supple, Barry. *The History of the British Coal Industry, IV:
1913–1946*. Clarendon Press, 1987.

WEBSITES

www.cmhrc.co.uk
(The Coalmining History Resource Centre, with a
searchable database of colliery accidents.)
www.dmm.org.uk
(For coal mining in the north of England.)
www.healeyhero.co.uk
(Covering mainly the East Midlands.)
www.scottishmining.co.uk
www.welshcoalmines.co.uk

INDEX

Page numbers in *italics* refer to illustrations

Aberfan 7
Accidents 24–7, 36, 40–3
Air shafts *16*, 23
Anderton shearer loader 32
Anthracite 9
Astley Green Colliery *28, 59*

Barnsley Main Colliery *4*
Bath, Temple of Minerva 10
Bell pits 11, *11*
Benwell Colliery *10*
Big Pit *19, 27*
Bituminous coal 8–9
Black damp 22
Brandling Main Colliery 24
Brass bands 57, *57, 58*
Brithdir 52
British Ironworks, Abersychan *23*

Canaries 44
Caphouse Colliery *60*
Chapels 56, *56*
Charter masters 46
Chatterley Whitfield Colliery *7, 33*
Children, employment of 36–8, *37, 38*
Choke damp 22
Clerk, Sir John 54–5
Clipstone Colliery 18
Coal-cutting machines *30*, 31–2, *31*, 45
Corves 13, 14, 29–30, *29*
Crumlin Navigation Pit *21*
Curr, John 13
Cwmbyrgwm Colliery *16*
Cwmgwrach, Neath 36

Davy, Sir Humphry 24
Denby Colliery *40*
Domesday Book 10
Drainage, *see* Mine Pumping
Dukeries 18, 54
Durham Miners Association 58:
    Durham Miners Gala 7, 58

Easington 55
Explosives 30

Farey, John 13
Felling Colliery 36
Ferndale Pit *43*
Fiennes, Celia 21
Firedamp 23, 24, 25, 44–5
Friendly Societies 46, 57

General Strike 47
Great Western Colliery *18*
Guibal fan 26

Hadrian's Wall 9–10
Harberlands Colliery 45
Hartley Colliery *42*, 52
Harworth Colliery 18
Heapstead *32*, 33, *33*
Hebburn Colliery 24, *32*
Hetton, County Durham 49
Hetton Silver Band *57*
Horse whim 14, *14*, 20, 21
Huskar Colliery 41

Inclined plane 15
Iron industry 5, 17

Jevons, W. Stanley 5

Leifchild, John 39–40, 53
Longwall mining 13–14

Manriding cars 44
Mecco-Moore cutter-loader 31
Mine pumping 21–2
Miners' Institutes *48*, 57–8
Mines rescue services 42–3, *44*
Murton Colliery 22

National Coal Board (NCB) 5, 18–19, 32
Newbold Moor, Leicestershire 21
Nixon's Navigation collieries 58
Norcroft Colliery 41
North Duffryn pit *26*

Oaks Colliery 26–7
Opencast mining 19

Penallta Colliery *28*
Penydarren collieries 24
Pillar-and-stall mining 12–13, *12*

Pit-head baths 43–4
Pit-head frames *4*, 28–9, *28*
Pit ponies 30, *45*
Pneumoconiosis 45
Pumping engines 22, *23*

Randall, John 38
Rhondda Valley *48*, 49, *50–1*, 55–6, *56*
Rosehall, Glasgow 53

Safety helmets 44
Safety lamps 24, *24*, 25–6, 44
Schiele fan *25*, 26
Screening 33
Senghenydd 7, 47
Slack coal 9
Spoil tips 5–6, 7, 30
Stephenson, George 24
Surgeons 46

Temple Main Colliery 35
Thoresby Colliery 18
Treorchy *48*
Trepanner 32
Tynewydd Colliery 42

Underground haulage 30, *30*
Unions 47, 57
    banners *57*

Ventilation 22–3, 26
Vobster Colliery 21

Waddle fan 26, *26*
Wages 45–6
Waggonways 14–15, *15*
Waldridge Colliery 52
Washery 33
Water balance *27*, 28
Wellsway Pit 41
Wideopen Colliery *6*
Winding 28–30, *29, 41*
Women, employment of 36, *37*, 38, *38*
Woodhorn Colliery *20*
Woodlands 54
Workington Coal Miners' Society 46